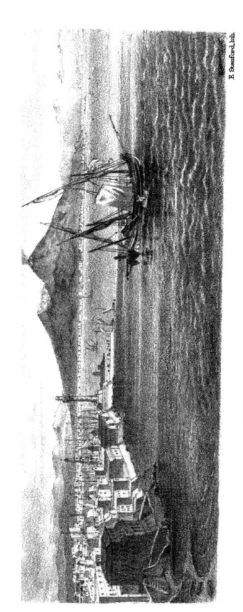

VESUVIUS FROM SANTA LUCIA, NAPLES.

MOUNT VESUVIUS:

A DESCRIPTIVE, HISTORICAL, AND GEOLOGICAL ACCOUNT OF THE VOLCANO,

WITH

A NOTICE OF THE RECENT ERUPTION,

AND

AN APPENDIX,

CONTAINING

LETTERS BY PLINY THE YOUNGER, A TABLE OF DATES OF ERUPTIONS, AND A LIST OF VESUVIAN MINERALS.

BY

J. LOGAN LOBLEY, F.G.S.

Illustrated with View, Map, and Section.

LONDON:
EDWARD STANFORD, 6 AND 7, CHARING CROSS, S.W.
1868.

PREFACE.

THE following description of Mount Vesuvius was prepared for the Geologists' Association of London, before which body it was read; but having been strongly urged to print it, and believing that a brief general and geological account of the volcano, with a notice of the eruption of this year, has not yet appeared, I venture to give these pages to the public.

The subject has been treated under four heads. In the first chapter, the geographical position of Vesuvius is described, and the various parts of the mountain indicated. In the second chapter is given the history of the volcano, with a *résumé* of the accounts which have come down to us of the most interesting and remarkable eruptions. The third chapter has been devoted to the geological and scientific consideration of Vesuvius. And in the fourth and last, an endeavour has been made to give a picture of the aspect which the mountain presents to the eye of the traveller who makes the ascent to the crater, and some account of the phenomena observed during the eruption of the present year.

The writings of Humboldt, Scrope, Daubeny, Forbes, Lyell, and Mallet, in our days, have left little to be said on the nature and origin of volcanic and earthquake phenomena, until fresh facts have been accumulated and given to the world. For information on this deeply interesting but theoretical branch of the subject, those well-known authors should be consulted. In the preparation of this descriptive account of Vesuvius, the narration of generally admitted facts and the results of personal observation, has been alone attempted; and the object contemplated by the writer will have been accomplished, if the following pages enable the unscientific, as well as the scientific, reader, to gather without difficulty, some useful information concerning one of earth's greatest wonders.

LONDON,
May, 1868.

CONTENTS.

	PAGE
PREFACE	v
GEOGRAPHICAL DESCRIPTION OF VESUVIUS	1
HISTORY OF VESUVIUS	6
GEOLOGY OF VESUVIUS	24
ASCENT OF VESUVIUS, AND ERUPTION OF 1868	34

APPENDIX.

EPISTLES OF PLINY THE YOUNGER, CONTAINING HIS ACCOUNT OF THE ERUPTION OF A.D. 79	45
TABLE SHOWING DATES OF ORDINARY AND PAROXYSMAL ERUPTIONS	53
LIST OF MINERALS FOUND ON VESUVIUS AND SOMMA	54

PLATES.

	PLATE
VIEW OF VESUVIUS FROM NAPLES (*Frontispiece*)	I.
MAP OF VESUVIUS	II.
GENERAL SECTION	III.

MOUNT VESUVIUS.

CHAPTER I.

INTRODUCTION, AND GEOGRAPHICAL DESCRIPTION OF VESUVIUS.

VESUVIUS, the celebrated volcano of Southern Italy, has been for many centuries an object of great interest to the inhabitants of Europe. In ancient times the conspicuous position of the mountain in one of the fairest and most frequented portions of the Roman dominions—the neighbourhood to which resorted the most wealthy, most famous, and most noble of the citizens of Rome—and the terrible character and dreadful results of the eruption of 79, combined to render Mount Vesuvius an object of especial interest and wonder. In modern times the proximity of the volcano to one of the greatest cities of Europe, the accessibility to travellers, and many attractions of the romantic coast of Italy, and the frequency and violence of the eruptions, have fixed the attention of mankind upon Vesuvius not less earnestly than in the days of old.

The mountain to which the general name of

Mount Vesuvius is given rises grandly to a height of upwards of 4000 feet on the north-eastern side of the Bay of Naples, contributing in no small degree to the world-famed beauty of its classic shores. Indeed the view of the bay would be without its distinctive character were this important feature of the scene wanting; and a pictorial representation of the city of Naples is at once recognized if the smoking cone of Vesuvius forms the background of the picture.

It may, perhaps, be well to at once point out that although Vesuvius is popularly considered to be a simple and an almost regular cone, yet the mountain has two peaks of nearly equal elevation. To one the name of Monte Somma is given, and to the other, which is the cone, having on its summit the active crater, the name of Vesuvius is more properly applicable. Monte Somma and the cone of Vesuvius are therefore the two peaks of the great mountain known as Mount Vesuvius, and which is at its base about thirty miles in circumference. This great base is nearly circular, having the apex of the great cone exactly over its centre, and the ridge of Monte Somma rising on its northern and eastern sides.

The whole mountain may therefore be described as consisting of a great circular base rising very gradually from the plains, with two comparatively small mountains standing upon it,—one a cone, over the centre of the whole, and the other a semicircular ridge, a little less in elevation than the cone which it half encircles.

The outline of the mountain, as seen from Naples, is on its southern side a singularly graceful curve, rising gently from the low grounds, and continued, with a gradually increasing inclination, to the summit of the cone, with, however, a slight elevation above the general curve, at a point about half-way to the summit, called La Pedamentina. On the northern side the outline corresponds with that on the south for a considerable distance from the plain, after which it increases rapidly in inclination to the summit of the ridge of Somma, which stands on the same level and at the same distance from the centre of the cone as the Pedamentina.

The deep semicircular valley lying between the ridge of Somma and the cone is called the Atrio del Cavallo at its northern end, and the Val del Inferno at its southern extremity. This valley has a level bottom, and extends for a distance of upwards of two miles at the foot of the cone, which rises on its other or seaward side from a comparatively level part of the mountain also. To this portion of the surface, stretching from the entrance of the Atrio del Cavallo to the Pedamentina, the name La Plaine has been given. Fronting the entrance to the Atrio del Cavallo, and with the Fosso dela Vetrana on one side and the Fosse Grande on the other, rises the Crocelle, the elevation on which stand the Hermitage and the Observatory.

On the southern side of the Fosse Grande, the Piano del Ginestre forms the terrace-like commencement of the plain at the foot of the cone.

Below the level of the Fosse Grande the sides of the mountain are cultivated, and the lower slopes gradually blend on the northern, eastern, and southern sides with the plains. On the south-western side the mountain rises almost from the water's edge, for although the shore of the bay is here lined with villages, or rather towns, they stand upon the lower slopes of the mountain itself, and are built upon the old lavas of the volcano.

Not only the western side, but the whole base of Vesuvius is skirted by a series of villages, some of which are very populous, especially those on the shore of the bay, where, although the eruptions of past times have frequently covered the ground with lava, and destroyed the habitations of man, an immense population is located, apparently not living in the slightest dread of the very dangerous neighbour whose threatening thunders they so often hear.

About six miles from the summit, in a direct line, and on the south-eastern side of the mountain, we find the ruins of Pompeii. At the base of the mountain, near the shore of the bay, and under the modern town of Resina, Herculaneum lies buried; while the ruins of Stabiæ, the third of the cities destroyed in the year 79, are close to Castelamare, at a distance of eight or nine miles from Vesuvius.

And here it may perhaps not be out of place to

remark, that an exceedingly fine view of the mountain, and, during the present eruption, of the red-hot lava on its side, may be obtained from various parts of the city of Naples, and especially from the Strada di Santa Lucia, where groups of people each evening congregate to witness the grand and striking scene.

CHAPTER II.

HISTORY OF VESUVIUS.

BEFORE proceeding to the consideration of the geology of Vesuvius, we shall do well to review briefly the history of this very interesting and remarkable volcano, since its internal structure and geological formation will be more readily understood when we have made ourselves somewhat acquainted with those eruptions which have largely contributed to the formation of the mountain as we now see it.

During the historic period previous to the Christian era Vesuvius was not an active volcano; and indeed it was not then generally known to be volcanic in origin, though several among the ancient naturalists suspected it to be so, and its structure, and the lithological character of the rocks of Somma, at once reveal to us the fact, that the mountain had its origin in volcanic action, and that although it had been dormant for many centuries, yet at a prehistoric period eruptions, perhaps more terrific than any which have occurred in modern times, had covered its sides with red-hot lava, and had ejected from a crater many times the size of the present one, ashes and cinders as now.

The form of the mountain was, however, during the period of its inaction, very different from that which in our times attracts the eye by its beautiful outline. Instead of a grandly swelling central cone, and a sharp and picturesque ridge standing on a widely spreading base, there was then only a truncated cone of great width and comparatively small height; for the present great cone of Vesuvius did not at that time exist, and the semicircular ridge of Somma of our day then extended quite round the area on which the cone now stands, forming a complete circle, and being the enclosing wall of the crater of the prehistoric volcano. This crater was no less than seven miles in circumference, with sides both lofty and precipitous. In the times of the early Romans the plain at the bottom of the crater was bare and sterile, while the rocks around were covered with wild vines extending from the top to the bottom.

Among the writers of antiquity who noticed the volcanic appearance of Vesuvius may be mentioned Diodorus Siculus and Strabo. The former of these ancient observers was born on the slope of a volcano, having been a native of Agyrium, at the base of Mount Ætna. And as two eruptions of Ætna occurred during his lifetime, he was well fitted to detect the signs of volcanic action. These signs he observed at Vesuvius, and thus has the honour of having been the first to notice the igneous and volcanic character of the mountain.* Following Diodorus Siculus, Vitruvius writes of Vesuvius, and mentions a tradition that fire had at one time been seen coming forth

---

* Diodorus Siculus, in his fourth book (Δ. 21), says:

'Ὠνομάσθαι δὲ καὶ τὸ πεδίον τοῦτο φλεγραῖον ἀπὸ τοῦ λόφου τοῦ τὸ παλαιὸν ἄπλετον πῦρ ἐκφυσῶντος, παραπλησίως τῇ κατὰ τὴν Σικελίαν Αἴτνῃ· καλεῖται δὲ νῦν ὁ τόπος Οὐεσούιος, ἔχων πολλὰ σημεῖα τοῦ κεκαῦσθαι κατὰ τοὺς ἀρχαίους χρόνους.

from the mountain. Strabo, however, like Diodorus Siculus, infers the volcanic origin of Vesuvius from the appearances presented by the rocks around the crater.* During the second Servile war, the crater supplied a place of refuge for Spartacus and his followers, who encamped within its bounds.

The mountain was therefore, at the commencement of the Christian era, an apparently extinct volcano, having a great general resemblance to many of those large volcanic hills which stud the Phlegræan fields on the western side of the city of Naples.

The year 63 of our era witnessed the earliest indications of renewed activity; and in the early part of that year, we are told by Seneca, that an earthquake occurred, which destroyed a considerable portion of both Herculaneum and Pompeii, and shook all the district surrounding the mountain. In the following year a second earthquake caused considerable injury to the city of Naples; and it is said that the theatre in which the emperor Nero had been performing a short time before was thrown down. These premonitory symptoms were followed, during

---

* In his fifth book of Geography (E. 4, 8) Strabo writes:

Νώλης δὲ καὶ Νουκερίας καὶ Ἀχερρῶν, ὁμωνύμου κατοικίας τῆς περὶ Κρέμωνα, ἐπίνειόν ἐστιν ἡ Πομπηία, παρὰ τῷ Σάρνῳ ποταμῷ καὶ δεχομένῳ τὰ φορτία καὶ ἐκπέμπουντι. ὑπέρκειται δὲ τῶν τόπων τούτων ὄρος τὸ Οὐέσουιον, ἀγροῖς περιοικούμενον παγκάλοις πλὴν τῆς κορυφῆς· αὕτη δ' ἐπίπεδος μὲν πολὺ μέρος ἐστίν, ἄκαρπος δ' ὅλη, ἐκ δὲ τῆς ὄψεως τεφρώδης, καὶ κοιλάδας φαίνει σηραγγώδεις πετρῶν αἰθαλωδῶν κατὰ τὴν χρόαν, ὡς ἂν ἐκβεβρωμένων ὑπὸ πυρός, ὡς τεκμαίροιτ' ἄν τις τὸ χωρίον τοῦτο καίεσθαι πρότερον καὶ ἔχειν κρατῆρας πυρός, σβεσθῆναι δ' ἐπὶ λιπούσης τῆς ὕλης. τάχα δὲ καὶ τῆς εὐκαρπίας τῆς κύκλῳ τοῦτ' αἴτιον, ὥσπερ ἐν τῇ Κατάνῃ, φασί, τὸ κατατεφρωθὲν μέρος ἐκ τῆς σποδοῦ τῆς ἀνενεχθείσης ὑπὸ τοῦ Αἰτναίου πυρὸς εὐάμπελον τὴν γῆν ἐποίησεν.

the reign of the emperor Titus, and in the year 79, by the famous eruption which not only destroyed three flourishing cities, but completely altered the form and appearance of Mount Vesuvius. As this eruption was such an important one, not only from the appalling results which followed, but also from its having been the first of the long series of eruptions which have interested and astonished the world during the historic period, I may perhaps be pardoned for directing the reader's attention to the letters* containing the celebrated account by Pliny the Younger, who was an eye-witness of the awful scene, and whose uncle was one of those who perished by the catastrophe.

It will be remarked that in this most interesting description there is no mention of lava having been seen; and this is quite in accordance with the results of modern observation, which has never yet detected any bed of lava of the same age as the ashes which overwhelmed Pompeii. It appears, therefore, that this great eruption was one entirely of ashes, lapilli, and mud, and that the melted rock which is given forth in modern times during nearly every eruption was then entirely wanting. From an examination of specimens brought from the ruins of the ancient city it will be found that the material which obscured the light of the sun as completely as Pliny describes, and buried the city of Pompeii and destroyed it for ever, was light, grey, pumaceous lapilli and ashes, evi-

* See Appendix, p. 45.

dently quite dry, from the fact that it is now in a perfectly loose and uncompacted state. The present condition of the numerous frescoes with which the houses of the wealthy inhabitants were adorned affords further evidence of this dryness, since the colours of these, in many cases, beautiful decorations are comparatively fresh and uninjured. It is probable, however, that the volcano discharged at a later period of the same eruption ashes and lapilli mingled with mud, since we find that the city of Herculaneum is buried beneath a mass of consolidated tuff, and not, as Pompeii, simply covered with loose ejectamenta. This great eruption left the mountain with only half of the walls of the ancient crater remaining, the sides on the south and west having been almost completely blown away.

It was now that the modern cone of Vesuvius began to rise and give that appearance to the whole with which we are familiar. Successive eruptions accumulated material on this new cone, and so added to its bulk and height, until it at length attained an elevation equal to that of the ancient summit; and thus we have the double-peaked mountain of the present century. There was, however, left on the southern side a small remnant of the wall of the old crater; and this, even to the present time, is seen as a protuberance on the side of the mountain, and is the elevation which was before mentioned under the name of the Pedamentina.

A second eruption, described by Dion Cassius,

HISTORY OF VESUVIUS.

occurred in the year 203, and one causing great destruction in 472. This latter eruption destroyed the villages which had been built over the buried cities of Herculaneum and Pompeii, and ejected ashes to so great a distance that it is said some fell as far from the mountain as Constantinople.

Many eruptions have occurred from that time to the present of more or less importance; but it will be sufficient to notice here merely those which are interesting from having modified the shape of the mountain or the character of its surroundings, and those during which some extraordinary phenomena were observed.*

The eruption of 1036 is important geologically, as the first-recorded fluid lavas are said to have flowed from the new cone in this year. This lava there is good reason for believing reached the sea.

During the next five hundred years volcanic action was very violent in the neighbouring districts. Several eruptions of Etna are recorded; the Solfatara, as well as the volcanic island of Ischia, poured forth streams of lava of great size; and the northern as well as the southern portions of Italy were convulsed by earthquakes. But during this period Vesuvius itself was comparatively inactive. This circumstance supports the view that all the volcanic vents of Italy have their origin in the same source, that the vent of Vesuvius affords in ordinary times a sufficient means of escape, or safety-valve as it

* See Appendix, p. 53.

were, in this district for the forces generated in the great interior reservoir, and that when this opening becomes sealed these forces open up other channels of communication with the exterior of the earth.

For a long time previous to the great eruption of 1631 the crater of Vesuvius contained so much vegetation that it became the resort of wild boars, and cattle grazed on the plain at the bottom. "Within the crater," writes Bracini, "was a narrow passage, through which by a winding path you could descend about a mile amongst rocks and stones, till you came to another more spacious plain covered with ashes. In this plain were three little pools placed in a triangular form,—one, towards the east, of hot water, corrosive and bitter beyond measure; another, towards the west, of water salter than that of the sea; the third, of hot water, that had no particular taste."

It was during this season of repose on the part of Vesuvius that occurred the extraordinary volcanic phenomenon of 1538. On the last three days of September in that year, Monte Nuovo, between Baiæ and Pozzuoli, was thrown up. This hill is about 440 feet in height, and encloses a very perfect crater, almost as deep as the mount is high. The formation of Monte Nuovo is especially interesting, as offering a probable explanation of the mode in which the lower portion of Vesuvius was originally formed, for since there was ocular demonstration of Monte Nuovo being a crater of eruption, there does not appear to be much difficulty in our supposing the great body of Vesuvius, as well as other elevations composed of volcanic matter, to have had a similar origin.

At the end of the year 1631 commenced one of the most important of the modern eruptions of the great Italian volcano. In this year the great crater became filled with volcanic matter, level with the brim, and on the 16th December a column of ashes mingled with vapour was ejected from the cone, while great quantities of stones were discharged and volcanic lightning flashed. After this no less than seven streams of lava descended the mountain, and an earthquake caused the sea to recede for half a mile. This great eruption resulted in a most awful destruction of human life, no fewer than 18,000 persons having been said to have perished during its continuance. The villages of Torre del Greco, Resina, Granatello, and Portici were either wholly or partially destroyed; and a large space of land was inundated by the torrents of rain produced by the condensation of the immense volumes of vapour discharged from the mountain. The lava emitted during this eruption forms a considerable portion of the substratum under the road between Resina and Torre del Annunziata, near which latter village I found it to be 20 feet thick, and reposing upon a bed of reddish-coloured volcanic ashes and cinders. It may be seen at many points along the high road, as well as by the side of the railway lying between the road and the sea, and it is quarried for road-material close to the seashore near Resina. It is this lava which covers the tuff under which Herculaneum lies buried, and which has rendered the excavations for the purpose of exploring the ruins of that ancient city so costly and so difficult; while above it stand the modern towns of Resina and Torre del Greco; and the beautiful gardens of the royal palace of Portici cover a portion of its surface.

An eruption, remarkable for having thrown up a perpendicular stream of lava, occurred in the year 1676; and during the next twenty years various eruptions took place, which much modified the shape of the mountain. At this period it appears that the highest point of Somma was 1200 feet higher than the summit of the new cone.

A stream of lava four miles in length, and upwards of 100 yards wide, where broadest, flowed from the mountain in 1694, and ran in the direction of San Giorgio a Cremano; and one, two years afterwards, towards Torre del Greco.

During the next hundred years the volcano was in very frequent activity; and several of the eruptions which occurred in this period are deserving of notice, on account of the remarkable nature of the phenomena witnessed during their continuance.

One of the early eruptions of the eighteenth century, that of 1707, was a very violent one; and is noticeable for the immense quantity of ashes then discharged by the volcano. It is stated that the cloud of ashes over Naples was dense enough to involve the city in such complete darkness that nothing could be seen in the streets. The shrieks of women, we are told, filled the air; and the churches were crowded with people, while the relics of S. Januarius were carried in procession. The wind, however, changing, and the ashes being carried in another direction, the alarm of the people subsided.

The eruption of 1737 poured forth a stream of lava no less than a mile in width, and containing upwards of 33,000,000 cubic feet. Sir Charles Lyell alludes to the lava of this eruption in his 'Principles of Geology,' in which he mentions that it may be seen near Torre del Greco, having a columnar structure. The section he speaks of is immediately outside the town of Torre del Greco, on the road to Torre del Annunziata, and may there be seen presenting a surface twelve or fifteen feet in height. This great stream flowed from the side of the mountain, but at the same time a stream was emitted from the crater on the summit, which divided into branches taking opposite directions, and causing great destruction to property. A great quantity of ashes was also ejected, and this ash falling on the trees in the surrounding district, especially in the neighbourhood of Ottaiano, also occasioned great damage. A curious mephitic vapour was emitted by the mountain after the eruption, which destroyed life.

In 1751, during an eruption which again produced a great flow of lava, the central cone sank, leaving a hollow of great depth and width.

Much lava was discharged in 1754, the stream

this year taking the direction of Bosco del Mauro and Bosco tre Case.

There was a remarkable outbreak in the year 1760. The lava of this eruption flowed from new cones which were formed, not on the summit of the mountain, but on its side, not far distant from the monastery of the Camaldoli. Great columns of smoke, and immense quantities of ashes, were also given forth by these new cones, which may still be seen.

The great eruptions of 1766, 1767, and 1770 are graphically described by Sir William Hamilton in his letters to the Royal Society, a short extract from which, having reference to the eruption of 1767, may not be out of place :—

"I observed, in my way to Naples, which was in less than two hours after I had left the mountain, that the lava had actually covered three miles of the very road through which we had retreated. It is astonishing that it should have run so fast; as I have since seen, that the river of lava, in the Atrio di Cavallo, was 60 and 70 feet deep, and in some places near two miles broad. When his Sicilian Majesty quitted Portici, the noise was greatly increased; and the concussion of the air from the explosions was so violent, that in the king's palace doors and windows were forced open; and even one door there, which was locked, was nevertheless burst open. At Naples, the same night, many windows and doors flew open; in my house, which is not on the side of the town next Vesuvius, I tried the experiment of unbolting my windows, when they flew open upon every explosion of the mountain. Besides these explosions, which were very frequent, there was a continued subterraneous and violent rumbling noise, which lasted this night about five hours. I have imagined that this extraordinary noise might be owing to the lava in the bowels of the mountain having met with a deposition of rain-water; and that the conflict between the fire and the water may, in some measure, account for so extraordinary a crackling and hissing noise. Padre Torre, who has wrote so much and so well upon the subject of Mount Vesuvius, is also of my opinion. And indeed it is natural to imagine that there may be rain-water lodged in many of the caverns of the mountain; as in the great eruption of Mount Vesuvius in 1631, it is well attested that several towns, among which Portici and Torre del Greco, were destroyed by a torrent of boiling water having burst out of the mountain with the lava, by which thousands of lives were lost. About four years ago, Mount Etna in Sicily threw up hot water also during an eruption.

"The confusion at Naples this night (19th October, 1767) cannot be described; his Sicilian Majesty's hasty retreat from Portici added to the alarm; all the churches were opened and filled; the streets were thronged with processions of saints: but I shall avoid entering upon a description of the various ceremonies that were performed in this capital to quell the fury of the turbulent mountain.

"Tuesday, the 20th, it was impossible to judge of the situation of Vesuvius, on account of the smoke and ashes, which covered it entirely, and spread over Naples also, the sun appearing as through a thick London fog, or a smoked glass; small ashes fell all this day at Naples. The lavas on both sides of the mountain ran violently; but there was little or no noise till about nine o'clock at night, when the same uncommon rumbling began again, accompanied with explosions as before, which lasted about four hours: it seemed as if the mountain would split in pieces. * * * * * During the confusion of this night, the prisoners in the public jail attempted to escape, having wounded the jailer, but were prevented by the troops. The mob also set fire to the Cardinal Archbishop's gate, because he refused to bring out the relics of Saint Januarius.

"Wednesday, 21st, was more quiet than the preceding days, though the lavas ran briskly. Portici was once in some danger, had not the lava taken a different course when it was only a mile and a half from it; towards night the lava slackened.

"Thursday, 22nd, about ten of the clock in the morning, the same thundering noise began again, but with more violence than the preceding days; the oldest men declared they had never heard the like; and, indeed, it was very alarming; we were in expectation every moment of some dire calamity. The ashes, or rather small cinders, showered down so fast, that the people in the streets were obliged to use umbrellas, or flap their hats, these ashes being very offensive to the eyes. The tops of the houses and the balconies were covered above an inch thick with these cinders. Ships at sea, twenty leagues from Naples, were also covered with them, to the great astonishment of the sailors. In the midst of these horrors, the mob, growing tumultuous and impatient, obliged the Cardinal to bring out the head of Saint Januarius, and go with it in procession to the Ponte Maddalena, at the extremity of Naples, towards Vesuvius; and it is well attested here, that the eruption ceased the moment the Saint came in sight of the mountain; it is true, the noise ceased about that time, after having lasted five hours, as it had done the preceding days."

An eruption of no great importance occurred in 1776; but three years after, in 1779, one of very great violence and remarkable character took place. Several great streams of lava were emitted on this occasion, and the showers of ashes were sufficiently dense to produce darkness in the vicinity of the mountain. Vapours most destructive both to animal and vegetable life were given off in profusion. But perhaps the most extraordinary feature of this terrible outbreak was the rising of a column of liquid fire, as it is described, to a height which Sir William Hamilton considered to be three times that of the mountain. After this a black cloud, emitting flashes of lightning, advanced from over Vesuvius towards Naples, spreading consternation and terror among the inhabitants of the city. The theatres were closed, and the relics of San Januarius again carried in procession. The destruction to property caused by this eruption was very great, especially to the vegetation of the district.

Terrible, however, as was the eruption of 1779, it was exceeded by the appalling outbreak of 1793-4, which lasted from the February of one year until the Midsummer of the next. This eruption produced no less than fifteen mouths, which emitted as many separate streams of lava; and these uniting formed one vast body of liquid fire, which flowed steadily on towards the sea, cutting in two the town of Torre del Greco, where it was from twelve to forty feet thick, and actually advancing into the sea to a distance of 362 feet, and presenting a face 1127 feet broad. It was calculated by Breislak that the whole mass contained no less than 46,098,766 cubic feet of lava. The sea was even in a boiling state at a distance of 100 yards from the lava. This eruption, as has been the case with the great eruptions generally, was preceded by a falling of the water in the wells of the neighbourhood.

From this time until 1822 the several eruptions which occurred were not of a very remarkable character; but on the 22nd of October, 1822, commenced one which must by no means be passed over. On the day following that I have named, the top of the cone fell in, and this was succeeded by the emission of a stream of lava nearly a mile in width. So great, too, was the quantity of ashes and cinders ejected, that the country as far as Amalfi, on the Gulf of Salerno, was overshadowed with darkness, and the road between Resina and Torre del Annunziata was blocked up with the fallen ejectamenta. This eruption was remarkable also for the unusually large amount of vapour which issued from the volcano. The vapour condensed into rain; and this was so great in quantity that some districts were inundated. The crater was greatly altered by this eruption, having been by it much increased in circumference and in depth, and the height of the cone correspondingly diminished. The summit was after this eruption estimated to be not more than 3400 feet above the level of the sea, 800 feet of the top of the cone having been blown away.

The eruption of 1828 produced a small cone in the centre of the large crater formed by the great outbreak of 1822; and in 1831 the summit of the new cone was higher than the edge of the great crater, which had become filled with the ejectamenta of the volcano. The whole of this new cone was, however, destroyed by the eruption of 1834, during which the volcano poured out a most copious stream of lava, causing the almost total destruction of the village of Caposecco, only four houses out of five hundred remaining. The river of lava which destroyed this village was nine miles long; and so great was its heat, that it is said to have been felt at Sorrento, a distance of nearly ten miles. One stream of the lava taking the direction of Pompeii, that ancient city was threatened with another interment.

Since 1834 the eruptions have been very frequent, and have been especially noticeable from the fact that they have produced lavas containing a greater quantity of leucite than is to be found in the lavas of the previous eruptions of the historic period, and therefore more like the ancient lavas of Monte Somma, in which this mineral is very abundant. We are told by Professor Pilla, that crystals of leucite as large as nuts were not only found imbedded in the lava of 1845, but were thrown out of the crater of the volcano in that year. This fact is interesting, as showing that these crystals of leucite were formed within the vent previous to the eruption.

The lava of 1850 formed a body no less than a mile and a half broad, which steadily advanced towards Bosco Reale and Ottaiano, consuming the woods of large trees which lay in its way, and threatening with total destruction the populous villages lying on that side of the mountain.

The great eruption of 1855 produced lava which retained its fluidity for such an unusually great length of time, that it flowed in a comparatively narrow stream almost to the suburbs of Naples, running between the villages of Massa di Somma and San Sebastiano, and reaching Cercola. This stream flowed from the side of the cone above the Atrio del Cavallo, into which it poured; and following the course of the Fosso della Vetrana and the Fosso

Faraone, it divided into two branches, one of which ran as above indicated to Cercola, the other took the direction of San Jorio, and almost reached that village. The lava, running so far into the cultivated region at the base of Vesuvius, caused a most deplorable destruction of property, and greatly terrified the inhabitants of the district, who feared it would destroy Portici, and perhaps reach even to the city of Naples.

Three years after, another eruption occurred; and it is the lava of 1858 which forms such a conspicuous feature on the side of the mountain when the ascent is made from Resina. This lava was emitted in copious streams from new craters, which opened in the Piano delle Ginestre, and flowed into the Fosse Grande at the base of the Crocelle.

The last great flow of lava previous to that of the present year occurred in 1861; but so low down on the mountain-side did the mouth open from which the stream poured, that it was not more than a mile distant from Torre del Greco. Besides the opening from which the lava was emitted, ten other craters were formed, which ejected great quantities of ashes. Torre del Greco again suffered greatly from this eruption, which caused the ground to be rent with fissures, and threatened to at length completely destroy that much-enduring city.

CHAPTER III.

GEOLOGY OF VESUVIUS.

BEING now acquainted with the various eruptions which have modified the form of the mountain, and gradually brought it to its present state, we shall, I think, have little difficulty in forming a tolerably correct judgment of its structure and geological formation.

We see that the entire mountain may be described as being made up of three distinct parts, each of a different age—the Great Cone, the ridge of Monte Somma, and the great base on which these stand.

It may be well, perhaps, to consider first the newest portion of the mountain, the Great Cone, and examine it by the light we have received from the history of its formation; and if we can obtain a clear idea of its structure, we shall have a knowledge of volcanic cones generally, since the cone of Vesuvius is quite a typical one, and may therefore be studied with great advantage.

This, perhaps, the most interesting part of the mountain, is at the present time a nearly regular cone, standing on a base about three miles in circumference, and having an elevation of about 1400 feet. The regularity and completeness of outline which is

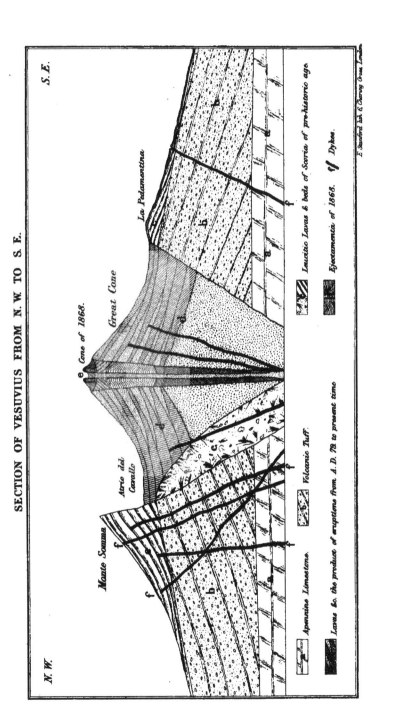

now seen has been produced by the present eruption, which has not only filled up the great crater, but on the plain so formed has raised a new cone with only a comparatively small crater. Thus at a short distance from the mountain little or no truncation is perceived; and the pleasing effect is produced of a beautifully proportioned conical mountain standing on an elevated base, and rising almost to a point.

The cone, being the result of successive accumulations of lava, ashes, cinders, and lapilli, ejected by a great number of eruptions, is composed of a series of layers, or coats, formed of these materials, and lying more or less parallel to the exterior slope, and therefore all having a quaquaversal dip from a central axis, which is the funnel, or neck, of the volcano, and through which the lava and vapours rise from the interior of the earth. The great enlargement of the crater in the year 1822 revealed the inner edges of the beds composing the cone, and the dip of the lava was ascertained. It was then found that these beds declined outwards, at angles varying from 26° to 30°. This, however, is not so great as the inclination of the exterior of the cone, which I found by careful measurement to be 40°. But we shall have no difficulty in accounting for this greater inclination, when we remember that the ashes and lapilli thrown out from the crater will lie thicker near the summit, and so tend to increase the inclination of the exterior surface.

As only a small proportion of the ejectamenta

discharged perpendicularly falls outside the circumference of the base of the cone, the beds in the Atrio del Cavallo are almost entirely formed of lava, and are there horizontal.

It was found when the examination above referred to took place that the layers of lava were crossed by vertical dykes of very compact basalt. These had doubtless been formed by the filling up of the channels through which the lava had at various times risen to the surface. The effect of these dykes or walls of basaltic rock will be to strengthen the whole structure of the cone, and so to enable it to withstand the rending and tearing force of the great eruptions.

On looking at the section, Plate III., we shall perceive how very simple and how very regular is the structure of a volcanic cone. It must not, however, be supposed that each layer or coat is carried round the cone, as sometimes there would be a deposit on one side and sometimes on another. The result, nevertheless, would give, were a section of the cone exposed, an appearance of regularity, as a layer extending only partially round the cone gradually becomes attenuated, and blends with the edge of the next which may be deposited on the contiguous portion of the exterior surface.

The mouth of the central volcanic vent and axis of the great cone, and to which the name of the crater is given from its cup-like form, varies in size and shape with almost every eruption. Sometimes it is a great yawning abyss of half a mile or more in diameter, and upwards of a thousand feet deep, as after the eruption of 1822, and sometimes it is a very small opening of not more than a hundred feet wide, as at the present time. After a great and, to to use Mr. Scrope's term, a paroxysmal eruption, the crater is very large. Minor eruptions following tend to fill up this great cup-shaped hollow by the accumulation of the ashes, stones, and lava which are given forth from the central vent. A cone is formed in the middle of the great crater, and this gradually becomes larger, until its apex is seen above the surrounding rim; thus the large crater is at last quite filled up, and a cone rises above it with a small crater at its apex. When this has occurred, the neck of the volcano is almost choked up, and so it remains, repressing as it were minor volcanic efforts, until the internal forces overcome the resistance, another great and paroxysmal eruption takes place, the whole of the top of the mountain is blown away, Vesuvius clears its throat, and we find again a vast and profound abyss descending into the very heart of the mountain.

It may, perhaps, be proper here to refer to the composition of the lavas of which the cone is to so large an extent made up, and which are chemically, mineralogically, and lithologically different from the ancient lavas of Monte Somma.

The modern lavas of Vesuvius, or those which have been emitted since the commencement of its renewed activity in the year 79, are remarkable for containing very little free leucite, and for yielding not more than six or seven species of minerals. These modern lavas are basalts or dolerites, being composed of augite and labradorite, with ferruginous matter, blended together in a compact and homogeneous mass, and forming a hard dark-coloured rock. It should, however, be noted that the lavas which have flowed from the volcano in deep and copious streams are very different in appearance from those which have issued in small streams, the former being much more compact than the latter, which, having cooled more rapidly, under little pressure, and with great exposure to the atmosphere, are very scoriaceous and spongiform in texture, and have an extremely harsh and trachytic feel.

Augite, titaniferous iron, hornblende, sodalite, mica, breislakite, and leucite are sometimes found in these lavas in detached crystals. Cotunnite was found after the eruption of 1822 inside the crater, and was also deposited in chinks in the lava of 1855 by sublimation. It is a chloride of lead, and contains upwards of 74 per cent. of the metal.

Monte Somma, the second of the three great divisions of the mountain, is, as has been previously shown, the remaining portion of the enclosing wall of the great and ancient crater which existed before the Christian era. The western and southern sides having been blown away, we find the ridge of Somma forming a semicircle, with its inner side facing the great cone on the north and east.

It has been calculated that the centre of the base of the modern cone is exactly coincident with the centre of the circle of which the escarpment of Somma forms a part. We thus see that the vent of the modern Vesuvius was also that of the pre-historic volcano, and that there has been no change in the position of the axis, such as has been the case in the island of Vulcano, off the coast of Sicily, where there are three eccentric craters.

The escarpment of Somma, extending for a distance of upwards of two miles, is more than a thousand feet high, and presents a face almost perpendicular. The upper edge is very irregular in outline, and is in fact a line of peaks of more or less angularity.

Unlike the great craters of the Campi Phlegræi, which are enclosed by cliffs of pumaceous tuff, the ancient crater of Vesuvius was encircled by walls formed of successive layers of lava, alternating with beds of scoriaceous and tuffaceous deposits, and traversed by dykes of basalt. The various layers of lava appear to be horizontal; but they are in reality not so—the appearance of horizontality being derived from the edges of the beds only being seen. These beds dip away from the exposed section at an angle of 26°, and so form the basis of the exterior slope of this part of the mountain, which has an inclination of about 30°.

The dykes crossing these beds are generally nearly perpendicular, though some may be seen rising at so low an angle as 45°. The basalt of which these dykes are composed is very similar to that of the beds of lava; but, as far as I could ascertain, is more homogeneous in texture.

The lavas of Somma are very hard and compact dolerites of a bluish grey colour, and contain, in addition to the augite and Labrador feldspar which form their essential constituents, the accessory mineral leucite in great abundance. Leucite is a silicate of alumina and potash, and is an opaque white mineral with a specific gravity of 2·48 and 5·5° of hardness. It crystallizes in the cubical or monometric system, and is often found here in detached and perfect trapezohedrons half-an-inch and sometimes more in diameter.

It is a remarkable and a suggestive fact, that although so few minerals have been found in the modern lavas of Vesuvius, upwards of 300 species have been enumerated as having been discovered in the ancient lavas of Somma. It should, however, be stated that the critical examination to which these minerals have been subjected by Professor Scacchi, of Naples, has resulted in reducing the number of distinct species, in the opinion of that eminent mineralogist, to about 40. The variety and abundance of the minerals found associated with the rocks of Somma, and their scarcity in the more recent productions of the volcano, very distinctly separate the ancient from the modern lavas. The minerals of Somma being so numerous, it would be out of place here to particularize them, and I will therefore only mention the more common species.*

Besides leucite, augite, hornblende, sodalite, and the other minerals found in the modern lavas, Somma yields idocrase in beautiful crystals, olivine, humite, nepheline, or sommite, garnets, apatite, aragonite, and glassy feldspar.

Underlying the lavas and beds of scoriæ and ashes of which the Great Cone and Somma are formed, we find the volcanic tuff of which the whole of the lower portion of the mountain is composed, and which is almost identical in mineral composition and lithological character with those vast accumulations forming the hills which lie around the city of Naples, as well as the elevations lining the bay on its northern side, extending from Naples to Misenum, and through which the great grotto of Posilipo is excavated. Of this tuff are also formed the encircling walls of the craters of the many extinct volcanoes of this extraordinary region; and it is evidently the oldest volcanic ejectamenta of the district. It has been well described by Dr. Daubeny, who says: "The basis of this tuff is for the most part of a straw-yellow colour, dull and hard to the feel, with an earthy fracture, and commonly a loose degree of consistency. It seems to be made up of comminuted portions of pumice, obsidian, trachyte, and many other varieties of compact, as well as of cellular lava,

* A complete list will be found in the Appendix, p. 54.

the softer kinds often rounded, the harder mostly angular. It is regularly stratified, and alternates with beds of loose uncemented pumice, of ferruginous sand, of loam, and in one or two cases of calc-sinter, in which fragments of limestone are impacted."

This tuff then forms the third, the oldest, and the largest of the great divisions of Mount Vesuvius, and can be seen as high up the mountain as the observatory which stands upon it, the whole ridge of the Crocelle being composed of this rock. It may also be seen on the lower part of Somma; but the western and southern sides of Vesuvius being so much encrusted with lavas from various eruptions, it does not appear on these, the most accessible parts of the mountain.

There can be no doubt that this mass of tufa, forming the great and spreading base of Vesuvius, was originally thrown out from a volcanic vent; and perhaps I may go further, and say that before the lava-formed ridge of Somma had any existence, it enclosed a great crater like those of the Phlegræan Fields. But into the vexed question, as to whether craters such as these are craters of elevation or of eruption, my limits will not permit me to enter. For, though I have no hesitation in saying I am strongly inclined to the opinion so ably advocated by Mr. Scrope and Sir Charles Lyell, and disposed to regard the raising of the great body of the mountain, as well as of Somma and of the cone, as being due to eruption, yet Professors Forbes and Daubeny, not to speak of foreign investigators, were such accurate observers and careful reasoners, and their opinion in favour of the elevatory theory consequently entitled to so much respect, that, in order to discuss the question fairly, the arguments on both sides must be very fully stated. The recent researches of our distinguished countrymen, Professors Phillips and Tyndall, will doubtless result in giving to us new and valuable data for the future discussion of this interesting question.

There cannot be, however, much doubt that by whichever way the lower portion of the mountain was raised, the tuff composing it was, in part at least, deposited under water; for, besides the fossil leaves which have been found, several genera of recent and tertiary marine mollusca have been taken from either this or the similar tuff of the neighbouring district. *Ostrea, Pecten,* and *Cardium* were found, according to Monticelli, in the tuff of Pozzuoli; *Ostrea* and *Pectunculus* are mentioned by Professor Pilla as having been taken from the similar rock at the castle of St. Elmo, Naples; and the tuff forming the promontory of Posilippo has yielded the genera *Ostrea, Cardium, Patella,* and *Buccinum.*

The whole mountain reposes on the fundamental rock of the district, the Apennine Limestone, which, since it contains *Hippurites,* may be considered to be of cretaceous age, and the probable equivalent of the chalk of the British Islands.

CHAPTER IV.

ASCENT OF VESUVIUS AND ERUPTION OF 1868.

HAVING considered Vesuvius geographically, historically, and geologically, let us now ascend the mountain and observe the appearances presented during the continuance of the present eruption.*

The ascent of Vesuvius is usually made from Resina, one of the towns near the shore of the bay, and about four miles from Naples. This town, built over the ancient city of Herculaneum, is on the great southern road from Naples, and has Portici with its royal palace on one side, and Torre del Greco, which has so repeatedly suffered from eruptions, on the other. The places I have named are rather parts of one long and populous town, and this again is connected with the city of Naples by continuous lines of houses and other buildings, forming a suburb of great size. The road, therefore, from Naples to the base of the mountain is lined with houses the whole distance, and is an exceedingly busy and noisy thoroughfare, abounding with scenes and objects calculated to arrest the attention of the lover of the picturesque, as well as of the student of national characteristics.

* During the continuance of the recent eruption a series of very interesting letters, giving a detailed account of the outbreak and of the changes which occurred from time to time in the phenomena observed, appeared in the pages of the 'Athenæum.'

As we ascend from the towns at the base and get clear of the houses, we find the sides of the mountain covered with vineyards and gardens in which the choicest fruit is grown. Indeed, the wine produced by these vineyards on the flanks of Vesuvius is famous for its quality, and so much is it esteemed by the Italians, that it has obtained the name of "*Lachryma Christi.*" The soil here is remarkably fertile, and is cultivated very highly. This great fertility is due to the feldspathic matter which it contains, and which has been derived from the decomposition of the underlying lava. The belt of cultivated land is about two miles broad, and this being passed, we reach the lava of 1858, and in a little time are in the Fosse Grande.

Here the scene which presents itself is in most striking contrast to that which gladdened the eye at the commencement of the ascent. In place of beautiful gardens in which the orange, the lemon, the almond, the fig, and the vine flourish in perfection, and in which roses and camellias bloom in profusion, we have around us a black sterile waste, covered with huge folds, waves, and unshapely masses of rough lava.

It is difficult to convey to those who have not seen a volcano a clear idea of the extraordinary aspect of the part of the mountain covered by the lavas of the last twenty years; but it may, perhaps, be imagined, if we can suppose a stormy sea of boiling pitch to have been suddenly cooled so as to retain, in a solid form, all the roughness, angularity, and irregularity which the surface had while liquid. In some places the lava has assumed the appearance of great coils of thick black ropes, in others sharp and rugged ridges rise, while in others again huge blisters cover the surface. This region of desolation is about a mile in breadth, and extends around the mountain at the base of the Great Cone and Monte Somma, which rise from it.

The path now ascends the Crocelle, on the summit of which stands the hermitage of S. Salvatore, where travellers may obtain rest and refreshment, and where there is a well of good water. The streams of lava which have at various times descended the part of the mountain near the Crocelle, have flowed at one side or other of this ridge, and so its summit has hitherto afforded a safe position for a habitation, while all around has been covered with a sea of liquid fire. Attached to the hermitage there is a sanctuary, with an altar and the shrine of the saint, whose remains lie beneath.

On account of the commanding position of this ridge, and its comparative immunity from danger, it has been selected as the site of an Observatory, built by the King of Naples, for the purpose of facilitating the observation and study of volcanic and earthquake phenomena. In this Observatory there is a collection of the minerals found around Vesuvius, which is shown to visitors by the intelligent and obliging custodian of the establishment. The Observatory is under the direction of Professor Palmiri, who has devoted great attention to Vesuvian phenomena.

I would strongly advise all who intend paying a visit to Vesuvius to at least ascend to the observatory, as it may be reached without difficulty, and the view which it commands gives a better idea of the whole mountain than can be obtained at, perhaps, any other point.

It is on that part of the mountain which lies under the eye of the observer while standing on the Crocelle and looking southwards that, during the present eruption, the streams of liquid lava may be found. At a little distance from these streams, in the daytime, no redness will be perceived; but the position of the hot lava is marked by the white fumes which rise from its surface. The atmosphere over a large extent may be seen glowing with heat; for the whole of this part of the mountain is covered with lava, which, though solidified, is still hot.

The lava of the present eruption has chiefly flowed from the summit of the Great Cone, and spread over the region at its base; but at the period of my visit, in the early part of March, there was no lava visible on the slope of the cone, as a crust had hardened round the stream, and so formed a tube through which the lava descended. On reaching the base of the cone, the fluid lava broke through the crust in many places, and was emitted in a great number of small streams. These streams rise from amongst the masses of hardened lava quite noiselessly, and flow slowly away towards the lower slopes. The lava, however, is of such great consistence, and cools so rapidly, that these small streams do not flow more than a hundred yards before they lose their fluidity, and break up, at a dull red heat, into cindery masses, which fall over the end of the ridge formed by the stream and roll down, becoming, on further cooling, of a dark-brown colour and vesicular and scoriaceous texture.

I have mentioned a ridge as being formed by the stream, and this is strictly correct; for although for a short distance from the source the lava cuts for itself a channel, yet by giving off those portions which, having come first into contact with the atmosphere, have cooled and solidified, it soon piles up a ridge, on the top of which it flows slowly along. This ridge, being nearly level, becomes as it lengthens higher and higher above the surrounding surface; and so at its termination it has a considerable elevation, and resembles very much the end of a railway embankment in process of formation. The solidified lava, on being pushed by the stream behind it to the end of this ridge, falls over, and so adds to its length. The rate at which the lava was flowing I found to be about 300 yards per hour, and the consistence so great that small stones thrown on to it would not make any impression or remain on its surface, and a considerable amount of force has to be used to detach a portion of the mass by means of a stick. Though the heat given off is very great, yet the stream may be approached quite closely and impressions of coins, &c., taken from it. Medallion portraits of Victor Emanuel and Garibaldi are produced, by means of moulds attached to long handles.

The neighbourhood of the hot lava is exceedingly rough and difficult to traverse, being covered with very sharp and variously-sized masses of partially-cooled lava, which scorch the boots and render very careful walking necessary.

At the commencement of the eruption, and again quite lately,* a magnificent stream of lava flowed down the exterior of the cone; and descending in a fiery cascade, and then dividing into two branches, it produced a very grand effect, especially when seen at night. Besides the lava which has been flowing during the last three months on the north side of the cone, a stream descended the opposite side of the mountain, and ran towards the woods attached to the casino of the Prince of Ottaiano. This stream has, however, only flowed occasionally, and not continuously like that on the north side.

From the Observatory the path continues along the ridge of the Crocelle, and then descends a little, after which it gently rises to the level of the Atrio del Cavallo, into which we now enter, having the steep and rugged sides of the great cone on our right hand, and the still more steep and rugged cliffs of Somma on our left. The scene is here desolate in the extreme. On either side and in front we see only the

* April, 1868.

dark and sterile surface of lava, entirely without verdure or vegetation of any kind, while we are hemmed in by vast and sombre walls forming a valley, fully justifying the name Val del Inferno given to a portion of this gloomy solitude.

We now commence the ascent of the cone, which is most difficult and laborious, for although the inclination is not more than 40°, the looseness of the very rough and angular masses covering the surface, and the consequent uncertainty of the footing, cause great fatigue. The assistance and encouraging exhortations of the guides, with constant effort on our part, enable us at length to reach the terrace at the summit of the old and the base of the new cone. We are here greatly excited by the proximity of the mouth of the volcano and by the deeply-interesting phenomena now seen; and spite of the warnings of the guides, we press on to ascend the new cone and gain the edge of the crater. This new cone is about 200 feet high, and has a comparatively smooth surface, with an inclination of about 30°.

The present small crater is oval in shape, with the northern lip much depressed; but its entire outline cannot be seen owing to the dense clouds of vapours which conceal the side towards which the wind may happen to blow. Besides the principal crater there is a smaller one on the side of the new cone, and a still smaller one on the terrace at the summit of the great cone. From these craters arise volumes of vapours charged with sulphurous fumes, and from the principal one stones and cinders of most irregular and various shapes are discharged with loud noises almost every minute. The stones and cinders rise to a great height, and if there be not much wind the greater number fall back again into the crater. The wind, however, usually carries a considerable proportion of the ejectamenta in its direction, and this falls on to the cone and so increases its bulk. Some of the stones, and these too of large size, fall at a considerable distance from and on the windward side of the crater; it is therefore scarcely safe to be near the crater, even when the side from which the wind may be blowing is chosen.

Great quantities of vapours rise from the sides of the new cone, especially near the top, and these fumes depositing sulphur and various salts, cover the surface with variously and beautifully-coloured incrustations.

On gaining the edge of the crater, and looking into its interior, a very curious and interesting scene is presented. Rolling clouds of dense white fumes are to be seen covering the whole of the bottom, and almost hiding the sides, while from the more distant part of the abyss the cinders and stones before mentioned are discharged. No flame or fluid lava was to be seen in these craters, but on looking down the smallest one the fumes were found to be illuminated, doubtless by the red-hot lava below.

The most interesting scene witnessed during the ascent is perhaps that seen from the top of the old cone, since at this point we are near to the crater, and we see the various volcanic phenomena with great distinctness, while we are startled by the loud noises produced in the interior of the mountain, and feel the tremor of the ground which the explosions occasion.

But the warnings of the guides must not be too much disregarded, and we prepare to leave the summit, though most reluctantly, for we cannot fail to experience the attractive power of the novelty and sublimity of the wondrous scene.

I cannot conclude this notice of Mount Vesuvius without a reference to the views to be obtained from its summit and its slopes. For the eye ranges over an expanse of sea and land which, for variety and picturesque outline of its general features, and for brightness and richness of colouring, is perhaps unsurpassed. The elegant curve of the shore line, extending from the promontory of Misenum to the base of the rugged heights above Sorrento, separates the blue waters of the Mediterranean from the vine-covered plain of the Campania. On our right the noble city of the Siren, with its long white arms, embraces as it were its own lovely bay, while the rocky isles of Ischia, Procida, and Capri stud the azure surface of the sea. On our left, and at our feet, lie the ruins of once rich and populous cities, which, with the other remains of ancient greatness with which the classic land of Italy abounds, give an additional charm to this most interesting, most remarkable, and singularly beautiful portion of the surface of our globe.

APPENDIX.

APPENDIX.

LETTERS OF PLINY THE YOUNGER,

CONTAINING AN ACCOUNT OF THE ERUPTION OF A.D. 79. FROM 'THE LETTERS OF PLINY THE YOUNGER.' BY JOHN EARL OF ORRERY.

PLINIUS CÆCILIUS SECUNDUS (CAIUS) *to* CORNELIUS TACITUS.
(Book VI., Epistle XVI.)

You are desirous that I should give you an account of the death of my uncle, that you may be enabled to transmit it to posterity with the greater truth. I return you thanks. I foresee that his death when celebrated by you must procure eternal honour to his name; for although his fall was attended by the destruction of most beautiful territories, seeming, as it were, destined to be remembered equally with those nations and cities who perish by some memorable event; although he had compiled works both numerous and lasting, yet the immortality of your writings will lengthen out the character which he has established to himself. I consider it as a blessing to be possessed of endowments which either qualify us for actions worthy of public record, or inspire us to write anything worthy of public attention. But I think those persons peculiarly favoured from heaven who obtain both these qualifications. My uncle by his own works and by yours may be numbered among these last, for which reason I more readily undertake, and even wish for, the employment that you enjoin.

He was at Misenum, where he had the command of a fleet which was stationed there. On the ninth of the calends of September, about the seventh hour, my mother informed him that a cloud appeared of unusual size and shape. After having reposed himself in the sun, and used the cold bath—he had tasted a slight repast, and was returned to his studies—he immediately called for his sandals, and repaired to a higher point of view, from whence he might more plainly discern this prodigy. The cloud (the spectators could not distinguish at a distance from what mountain it arose, but it was afterwards found to be Vesuvius) advanced in height; nor can I give you a more just representation of it than the form of a pine-tree, for springing up in a direct line, like a tall trunk, the branches were widely distended. I believe, while the vapour was fresh, it more easily ascended; but when that vapour was wasted the cloud became loose, or, perhaps oppressed by its own gravity, dilated itself into a greater breadth. It sometimes appeared bright, and sometimes black, or spotted, according to the quantities of earth and ashes mixed with it. This was a surprising circumstance, and it deserved, in the opinion of that learned man, to be inquired into more exactly. He commanded a *Liburnian* galley to be prepared for him, and made me an offer of accompanying him if I pleased. I replied it was more agreeable to me to pursue my studies; and, as it happened, he had allotted me something at that time to write. He went out of the house with his tablets in his hand. The mariners at Retina being under consternation at the approaching danger (for that village was situated under the mountain, nor were there any means of escaping but by sea), entreated him not to venture upon so hazardous an enterprise. He continued firm to his resolution, and performed, with great fortitude of mind, what he had at first undertaken from a thirst of knowledge.

He commanded the galleys to put off from land, and embarked with a design not only to relieve the people of Retinæ but many others in distress, as the shore was interspersed with a variety of pleasant villages. He sailed immediately to places which were abandoned by other people, and boldly held his course in the face of danger, so composed as to remark distinctly the appearance and progress of this dreadful calamity and to digest and dictate those remarks.

He now found that the ashes beat into the ships much hotter and in greater quantities; and as he drew nearer, pumice-stones, with black flints, burnt and torn up by the flames, broke in upon them. And now the hasty ebb of the sea, and ruins tumbling from the mountain, hindered their nearer approach to the shore. Pausing a little upon this, whether he should not return back, and instigated to it by the pilot, he cries out, "*Fortune assists the brave: let us make the best of our way to* POMPONIANUS," who was then at Stabiæ, and lay opposite to a bay into which the sea, creeping gently along that winding coast, insinuates itself. Pomponianus, although not in immediate peril, yet seeing it plainly and finding it approaching fast, was putting his baggage on board some vessels, with a design of making his escape by sea whenever the contrary wind should abate. My uncle, arriving with a fair wind at this place, embraced, comforted, and encouraged his trembling friend; and to effect this, seemed himself to be under no kind of apprehension, but ordering his servants to carry him to the bath, when he had bathed, went to supper, either with a real cheerfulness, or, what is equally the sign of a great mind, the appearance of it.

In the meantime flames issued from various parts of Mount *Vesuvius*, and spreading wide and towering to a great height, made a vast blaze, the glare and horror of which were still increased by the gloominess of the night.

My uncle, to move the general fear, said that the blaze was occasioned by the villages being on fire which were now deserted by the country people. Then retiring to take his rest, he enjoyed a sound sleep; for being of a gross and corpulent habit of body, he was heard to snore by those who waited upon him. The court beyond which was his apartment by this time was so filled with cinders and pumice-stones that had he continued any longer in his room his passage from it would have been stopped up. Being awakened therefore, he quitted his chamber and returned to Pomponianus and the rest, whose fears had hindered them from sleeping, and who had been upon the watch. They consulted together whether it would be more advisable to keep under the shelter of that roof or retire into the fields; for the house tottered to and fro, as if it had been shaken from the foundation, by the frequent earthquakes. On the other hand, they dreaded the stones, which, by being burnt into cinders, although they fell with no great weight, yet fell in large quantities. But after considering the different hazards which they run, the advice of going out prevailed. In others, one kind of fear conquered another; in my uncle, one prudential reason only succeeded to another.

They covered their heads with pillows bound with napkins: this was their only defence against the shower of stones. And now, when it was day everywhere else, they were surrounded with darkness blacker and more dismal than night, which however was sometimes dispersed by several flashes and eruptions from the mountain. They agreed to go farther in upon the shore, and to look out from the neighbouring land if they might venture to sea; but the sea continued raging and tempestuous. Then my uncle, laying himself down upon a cloth spread on the ground, called twice for some water, and drank it; but the flames and a stench of sulphur which preceded them obliged others to immediate flight and roused him. He raised himself upon his feet, supported by two servants; but his respiration being stopped, he immediately dropped down, stifled, as I imagine, by the sulphur and grossness of the air. His lungs, as he was narrow chested, were naturally weak and subject to inflammations. When the light returned, which was not till the third day after his death, his body was discovered, untouched by the fire, without any visible hurt, in the dress in which he fell, appearing rather like a person sleeping than like one who was dead.

My mother and I still continued at *Misenum*. But this has no relation to the history, nor did you desire any particulars except those of my uncle's death. I shall therefore finish my letter, adding only that I have sent you all the circumstances which I either saw myself or were communicated to me at a time when the truth of every single incident could be easily recollected. From hence you will select such passages as you shall think proper; for it is one thing to write a letter, another to compile a history: nor is the difference less between writing to a friend in particular than to the world in general. Farewell.

PLINIUS CÆCILIUS SECUNDUS (CAIUS) *to* CORNELIUS TACITUS.
(Book VI., Epistle XX.)

You tell me that my former letter, which at your own desire I wrote to you concerning my uncle's death, has tempted you to inquire not only into the terrors, but the distress I suffered while I was left at *Misenum,* for with that particular my letter concluded.

"I will restrain my tears, and briefly tell." When my uncle was gone from us I employed my time (having stayed behind for that purpose) at my studies. I bathed, went to supper, and had a very imperfect and restless sleep. We had for several preceding days together felt an earthquake, which, being common in *Campania,* did not much alarm us; but the shocks were so violent this particular night that all things around us were not only moved, but seemed upon the brink of destruction. My mother hastened into my bed-chamber, at the moment of time when I was rising with an intention to awaken her if I had found her sleeping. We retired into a little court which lay between the house and the sea. I am in doubt whether my conduct ought to be called fortitude or thoughtlessness upon this occasion, for I was then but eighteen years of age. I called for a "*Livy,*" and read it as if I had been quite at ease; and, in the manner I had begun, went so far as to select passages from that author.

A friend of my uncle's, who was lately come from *Spain* on purpose to see him, finding my mother and me sitting thus together, and taking notice that I was reading, reproved the patience of her temper and the indifference of mine. However, I still continued intent upon my book. It was now six o'clock in the morning, yet there was but a faint and glimmering light. The house shook violently; and though we were in an open court, yet, as it was very narrow and built almost all round, we were certainly in great danger. We then thought it expedient to leave the town : the people, distracted with fears, followed us, and (such is the nature of fear which embraces, as most prudential, any other dictate in preference to its own) they pressed upon us and drove us forward. When we were out of reach of the buildings we stopped ; our astonishment was great, nor were our apprehensions less, for the carriages which we had ordered out of the town were so violently shaken from side to side, although upon plain ground, that they could not be kept in their places even when propped by heavy stones. The sea, too, seemed to be forced back upon itself, repelled as it were by the strong concussions of the earth. It is certain that the shore was greatly widened, and many sea-animals were left upon the strand.

On the land side a dark and horrible cloud, charged with combustible matter, suddenly broke and shot forth a long trail of fire, in the nature of lightning, but in larger flashes. Then my uncle's friend, the same who came out of *Spain*, said to us, with great vehemence and eagerness, " If your brother and your uncle be still living, his wishes are employed for your safety. If he has lost his life, he was desirous yours might be saved. Why then will you not immediately leave this place ?" We answered that we were not so solicitous for our own as for my uncle's preservation. He then hastily withdrew, running with the utmost expedition from danger. Not long after, the cloud descending covered the whole bay, and we could no longer see the island of *Caprea* or the promontory of *Misenum*. My mother now began to beseech, advise, and command me to make my escape in any manner I could. She observed that as I was young I might easily take my flight; but that she, who was in years and less active, could patiently resign herself to death, in case she was not the occasion of my destruction. My answer was, "I will never attempt at safety if we are not together." And then, leading her by the hand, I assisted her to go faster; she yielded with regret, still angry at herself for delaying me.

The ashes now fell upon us; however, in no great quantities. I looked back. A thick dark vapour just behind us rolled along the ground like a torrent, and followed us. I then said, "Let us turn out of this road, whilst we can see our way, lest the people who crowd after us trample us to death." We had scarce considered what was to be done, when we were surrounded with darkness, not like the darkness of a cloudy night or when the moon disappears, but such as is in a close room when all light is excluded. You might then have heard the shrieks of women, the moans of infants, and the outcries of men. Some were calling for their parents, some for their children, some for their wives: their voices only made them known to each other. Some bewailed their own fate; others the fate of their relations. There were some who, even from a fear of death, prayed to die. Many paid their adorations to the gods; but the greater number were of opinion that the gods no longer existed, and that this night was the final and eternal period of the world. There were others who magnified the real dangers by imaginary and false terrors. Some affirmed that *Misenum* was burnt to the ground. The report, although not true, gained credit.

A little gleam of light now appeared. It was not daylight, but a forewarning of the approach of some fiery vapour—which, however, discharged itself at a distance from us. Darkness immediately succeeded. Then ashes poured down upon us in large quantities, and heavy, which obliged us frequently to rise and brush them off, otherwise we had been smothered or pressed to death by their weight.

I might boast that not one sigh or timorous word broke from me through all this distress, had I not fortified myself with one great consolation—a miserable one indeed—that all nature was perishing with me.

At last this darkness, which now was drawn into the thinness of a cloud or of smoke, went off; true day appeared. The sun shone forth, but pale, as at the time of an eclipse. All objects that offered themselves to our sight (which was yet so weak that we could scarce bear the return of light) were changed, and covered with ashes as thick as snow. At our return to *Misenum*, after having refreshed ourselves, we remained in that suspense and doubt of mind which hope and fear inspire: fear indeed was most prevalent; for the earthquake still continued, and several enthusiasts, by dreadful prophecies, increased their own fears and the fears of others. But although we had undergone many dangers, and dreaded still more, yet we could not be persuaded to quit the town till we had received some intelligence concerning my uncle.

You will read this account without any intention of making it a part of your history, of which it is by no means worthy; and you must blame yourself for requiring it from me, if you think it not worthy of a letter. Farewell.

ERUPTIONS OF VESUVIUS.

From the renewal of the activity of Vesuvius, A.D. 79, to the present time, sixty-six eruptions have been recorded, of which fifty-three were of an ordinary character, the remaining thirteen being considered by Mr. Scrope to have been *great* or paroxysmal eruptions. The eruption which excavated the great crater of the ancient mountain, and which occurred in prehistoric times, must also have been paroxysmal and of terrific force and violence, as well as, very probably, of long duration.

The following Table gives the dates of the various historic

APPENDIX. 53

eruptions of the volcano, asterisks indicating the years in which paroxysmal eruptions took place.

A.D.
*79. Pompeii, Herculaneum, and Stabia destroyed.
*203.
*472.
*512.
*685.
983.
*993.
*1036. First recorded fluid lava.
1049.
*1138 or 9. 6 Kal. Jan.
*1306.
1500.
*1631. Great flow of lava—18,000 persons perished.
1660.
1676. Perpendicular column of lava.
1682.
1694. March 12, with feeble recurrences till 1698.
1698.
1701.
1707. May 20, with feeble recurrences till August.
1712. February 18, eruption continued till following year.
1717. June 6, continued as before.
1720. Eruption of ashes without lava.
1727. July 26.
1730. February 27.
1737. May 14, great flow of lava.
1751. October 25.
1754. December 2.
1758.
*1760. December 23, new crater on mountain side opened.
1766. March 25.
1767. October 23.

A.D.
1770.
1776. September 22.
1779. August 3, perpendicular column of lava.
1783. August 18.
1784. October 12 and December.
1786. October 31.
1787. December 21.
1788. July 19.
1789. September 6.
*1794. June 15, great flow of lava, Torre del Greco suffered great injury.
1799. February.
1804. August 12 and November 22.
1805. July.
1806. May.
1809. December 10.
1811. October 12.
1811. December 31.
1813. May to December.
1817. December 22 to 26.
1818.
1819. April 17.
1819. November 25.
1820.
*1822. 800 feet of the cone blown away; great discharge of ash and vapour.
1828.
1831. August 18.
1834. August.
1838. Minor eruptions of scoria.
1845. Ditto.
1850.
1854.
1855. Lava very fluid.
1861.
1868.

APPENDIX.

LIST OF MINERALS FOUND ABOUT VESUVIUS AND SOMMA, THE NAMES OF THOSE ONLY BEING GIVEN WHICH ARE CONSIDERED BY PROFESSOR SCACCHI TO BE SPECIFICALLY DISTINCT.

WHERE FOUND.	NAME OF MINERAL.	SYNONYMS.	CHEMICAL COMPOSITION.	CRYSTALLOGRAPHIC SYSTEM.
Somma	Analcime	Kubizit (Zeolite)	Silicate of soda + silicate of alumina + water	Monometric.
,,	Anorthite	Christianite, Biotina	Silicate of lime + silicate of alumina	Triclinic.
,,	Apatite	Moroxite	Phosphate of lime	Rhombohedral.
,,	Aragonite	Needle spar	Carbonate of lime	Trimetric.
Somma and Vesuvius	Augite		Silica + lime + magnesia + alumina + protoxide of iron	Monoclinic.
Somma	Blende	Sphalerite	Sulphuret of zinc	Monometric.
Vesuvius	Breislakite			
Somma	Calcite		Carbonate of lime	Rhombohedral.
,,	Calcopyrite		Sulphuret of copper and iron	Dimetric.
,,	Comptonite	Thompsonite (Zeolite)	Silicate of lime and soda + silicate of alumina + water	Trimetric.
,,	Copper pyrites		Sulphuret of copper and iron	Dimetric.
Vesuvius	Cotunnite	Cotunnia	Chloride of lead	Trimetric.
Somma	Dayrne		Silica + alumina + lime + potash	Rhombohedral.
,,	Feldspar (glassy)		Silicate of alumina + potash	Monoclinic.
,,	Fluor spar		Fluoride of calcium	Monometric.
,,	Galena		Sulphuret of lead	,,
,,	Garnet		Silicate of alumina + lime + peroxide of iron + manganese	,,
,,	Gismondine	Zeagonite (Zeolite)	Silica + alumina + lime + potash + water	Trimetric.
,,	Göthite	Hydrate of iron	Peroxide of iron + water	
,,	Hauyne	Berzein	Silicate of alumina and soda + sulphate of lime	Monometric.
Somma and Vesuvius	Hornblende	Amphibole	(Silica + magnesia + lime + alumina + protoxide of iron + manganese + hydrofluoric acid and water	Monoclinic.
Somma	Humboldtilite	Zurlite	(Silicate of alumina and lime + magnesia + soda + potash + peroxide of iron	Dimetric.

APPENDIX.

Locality	Mineral	Composition	System
Somma	Humite	Silicate of magnesia + protoxide of iron + fluorine	Trimetric (Scacchi). Monoclinic (Phillips).
,,	Idocrase	Silicate of alumina + silicate of lime + magnesia + protoxide of iron + protoxide of manganese	Dimetric.
,,	Iron pyrites	Bisulphuret of iron	Monometric.
,,	Lapis lazuli	Silicate of alumina + sulphate of soda + lime + traces of iron, chlorine, and water	,,
Somma and Vesuvius	Leucite	Silicate of alumina + silicate of potash	,,
Somma	Magnesite	Carbonate of magnesia	Rhombohedral.
Vesuvius	Magnetite	Peroxide of iron + protoxide of iron	Monometric.
Somma	Meionite	Silicate of alumina and lime	Dimetric.
,,	Mellilite	Silicate of alumina + lime + magnesia + peroxide of iron	,,
Somma and Vesuvius	Mica	Muscovite, Biotite	Silicate of alumina and potash + magnesia + peroxide of iron + fluoric acid + water	Trimetric.
Somma	Micaceous iron	Specular iron	Peroxide of iron + titanic acid	Rhombohedral.
,,	Nigrine	Rutile	Oxide of titanium + peroxide of iron	Dimetric.
,,	Nepheline	Sommite	Silica + alumina + soda + potash + lime + iron and water	Rhombohedral.
,,	Olivine	{Peridote, Monticellite, Chrysolite}	Silicate of magnesia + protoxide of iron	Trimetric.
Furnaroles	Pleonaste	Spinel	Alumina + magnesia + protoxide of iron + silica	Monometric.
,,	Realgar		Sulphuret of arsenic	Monoclinic.
,,	Sarcolite	Analcime carnea	Silicate of alumina + soda + lime	Dimetric.
Vesuvius	Sodalite		Silicate of alumina and soda + chloride of sodium	Monometric.
Somma	Sphene	Titanite	Silicate of lime + titanic acid	Monoclinic.
,,	Wollastonite	Tabular spar	Silicate of lime	Monoclinic.
,,	Zircon	Zirconia	Silicate of zirconia	Dimetric.